SPONGES

Rebecca Woodbury, Ph.D., M.Ed.

Gravitas Publications Inc.

Sponges

Illustrations: Janet Moneymaker

Sponges
ISBN 978-1-950415-57-1

Published by Gravitas Publications Inc.
Imprint: Real Science-4-Kids
www.gravitaspublications.com
www.realscience4kids.com

Sponges are squishy animals.

Do you think we are squishy?

No.

All sponges live in water.

Most sponges live in the ocean.

Some sponges live in fresh water.

Some sponges are small
and stick to rocks.

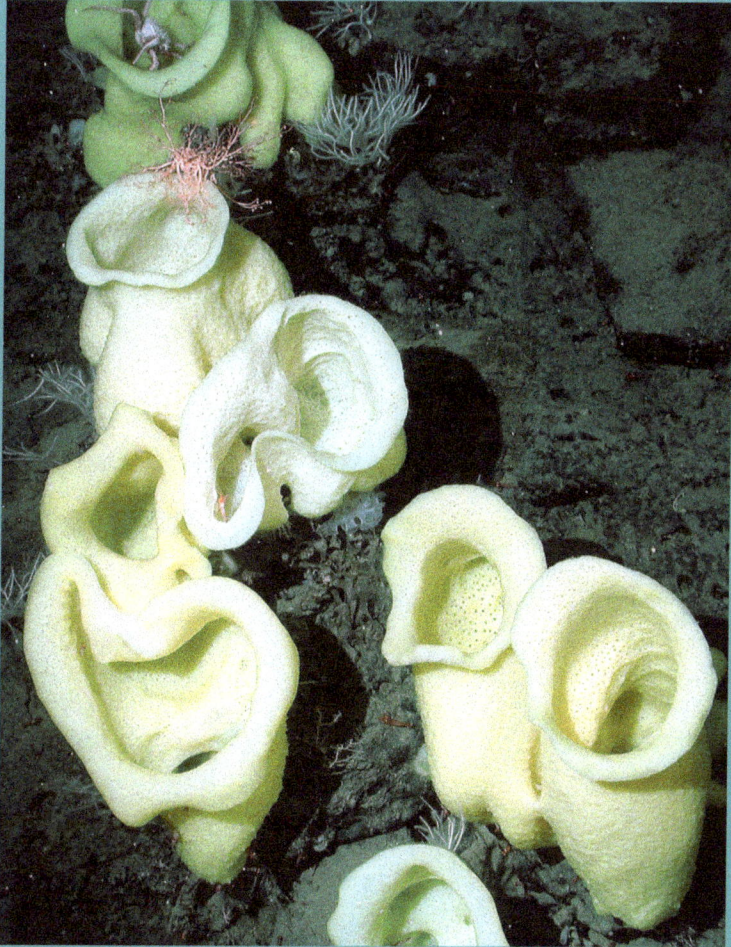

Some sponges are very BIG!

Look! A diver is inside that sponge!

Sponges come in many different colors.

Sponges come in many different shapes.

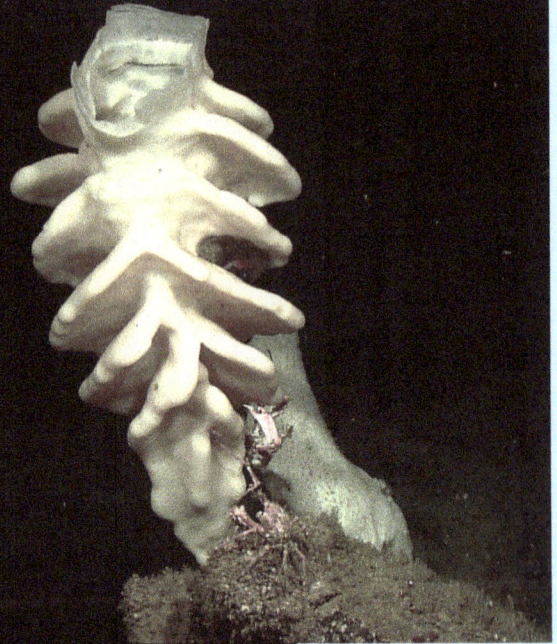

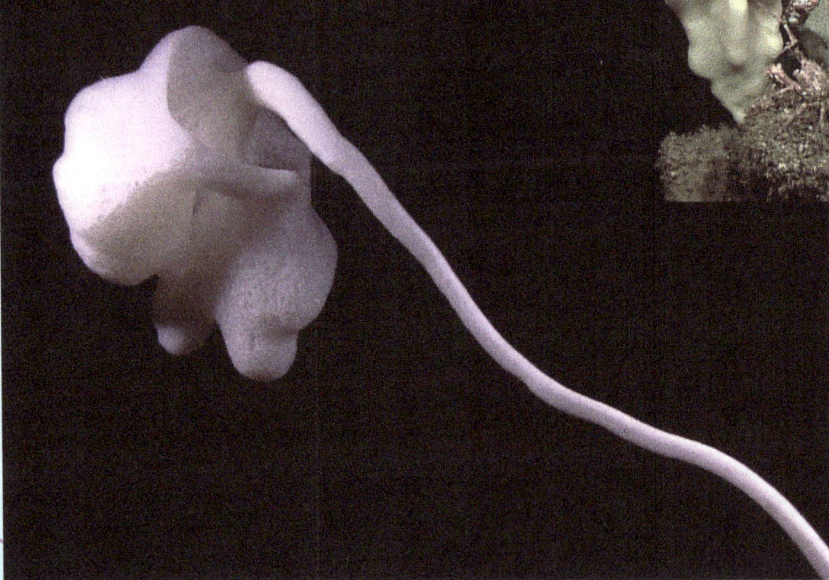

Sponges are sometimes hiding places for fish.

Some sponges may look like plants...

...but sponges are not plants.

That looks like it could be a plant.

But what are sponges really?

Turn the page to find out!

Sponges are animals.

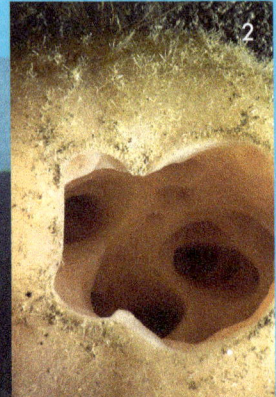

Sponges are animals

BECAUSE

sponges have **animal cells.**

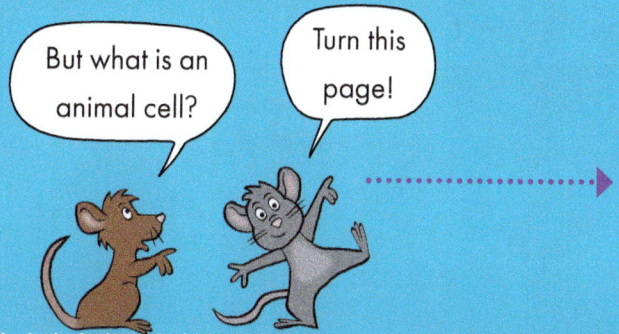

Review

An **animal cell** is a type of cell found only in animals.

Review

Living things are made of **cells.** **Cells** do lots of jobs inside living things. **Cells** are made of **atoms** and **molecules.**

Review

Molecules are made when **atoms link** together.

Review

Atoms are tiny building blocks that can link together.

Atoms make everything we see, touch, taste, and smell.

SPONGE FACTS!

- Sponges are animals that live in water. Most live in the ocean. Some live in fresh water.

- Sponges are made of animal cells.

- Sponges can be big or small and come in different colors and shapes.

How to say science words

animal cell   (AN-uh-muhl SEL)

atom   (AA-tuhm)

cell   (SEL)

molecule   (MAH-luh-kyool)

ocean   (OH-shuhn)

science   (SIY-uhns)

sponge   (SPUNJ)

www.ingramcontent.com/pod-product-compliance
Lightning Source LLC
Chambersburg PA
CBHW041429270326
41933CB00023B/3491